FABULA NUMERORUM

THE NUMBER STORY

SMALL BOOK ONE

ENGLISH - LATIN

Numbers Teach Children
Their Number Names

written and illustrated by

MISS ANNA

Early Reader Edition of *The Number Story 1*
Bronze Medal Winner, 2016 Wishing Shelf Book Award

Library of Congress Control Number: 2018902040

Names: Miss Anna, author.
Title: Number story : numbers teach children their number names / Miss Anna.
Description: Portland, OR: Lumpy Publishing, 2018.
Identifiers: ISBN 978-1-949320-11-4 | LCCN 2018902040
Summary: The pictures and rhymes present stories which introduce numbers 0-10.
Subjects: LCSH Numeration—English—Latin--Pictorial works--Juvenile literature. | BISAC JUVENILE NONFICTION /
Languages: English—Latin
Classification: LCC QA141.3 .M57 2018 | DDC 513—dc23

Publisher: Lumpy Publishing
Website: www.missannabooks.com
Email: missanna@missannabooks.com

Paperback: ISBN 978-1-949320-11-4
Printed in the U.S.A.   1 3 5 7 9 10 8 6 4 2

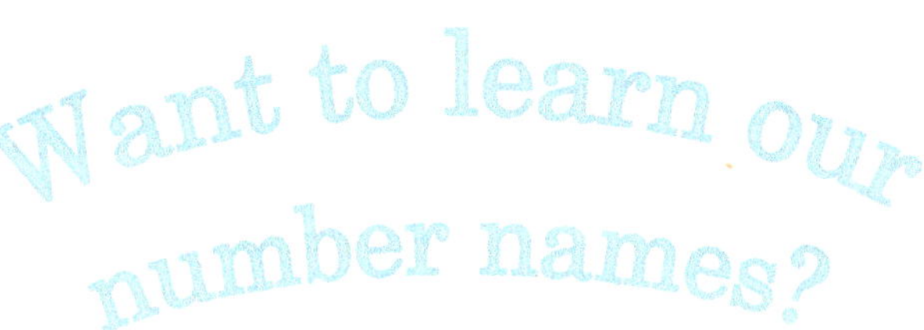

Nonne vis nomina numeorrum discere?

It is very easy and a lot of fun!

Id est facillimum et iucundissimum!

Say-along our little jingle

Cane nobiscum nostram fabulam brevem!

starting from Number One!

A Numero Uno incipiemus!

ONE looks like my one finger.

I ☆ UNUS

similis digito uno meo est.

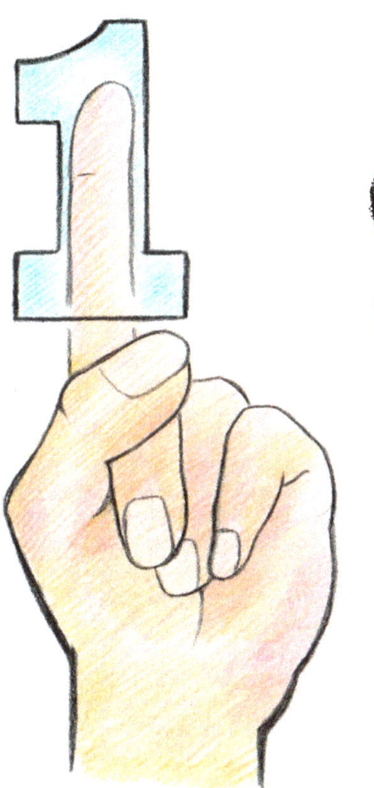

ONE!

UNUS!

2

TWO trails a tail.

II ☆ DUO

Caudam capit.

A TAIL! CAUDA!

3

THREE   has bumps.

III ☆ TRES

Montes habet.

BUMPY! MONTUOSI!

4

FOUR carries a sail.

IV ☆ QUATTUOR

Vela vehit.

4
A SAIL!
VELA!

5

FIVE   is a racing track.

V ✫ QUINQUE

Circenses est.

VROOM
1

6

SIX curves like a snail.

VI  SEX

Ut cochlea curvat.

A SNAIL! COCHLEA!

7

SEVEN has a sharp angle.

VII ✦ SEPTEM

Angulum acutum habet.

BE CAREFUL! IT'S SHARP!

CAVE! ACUTUM EST!

8

EIGHT is rollercoaster rails.

VIII ★ OCTO

Mons Russicus est.

EUGE!
YIPPEE!

9

NINE is a bubble on a stick.

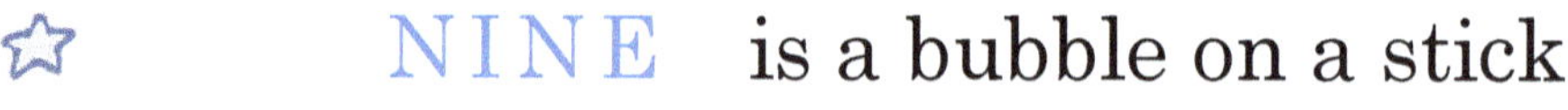

IX ✶ NOVEM

Bulla in baculo est.

A BUBBLE!

BULLA!

TEN is an eye of a whale.

X ☆ DECEM

Oculus unus balaenae est.

WINK!

CONIVE!

HELLO! SALVE!

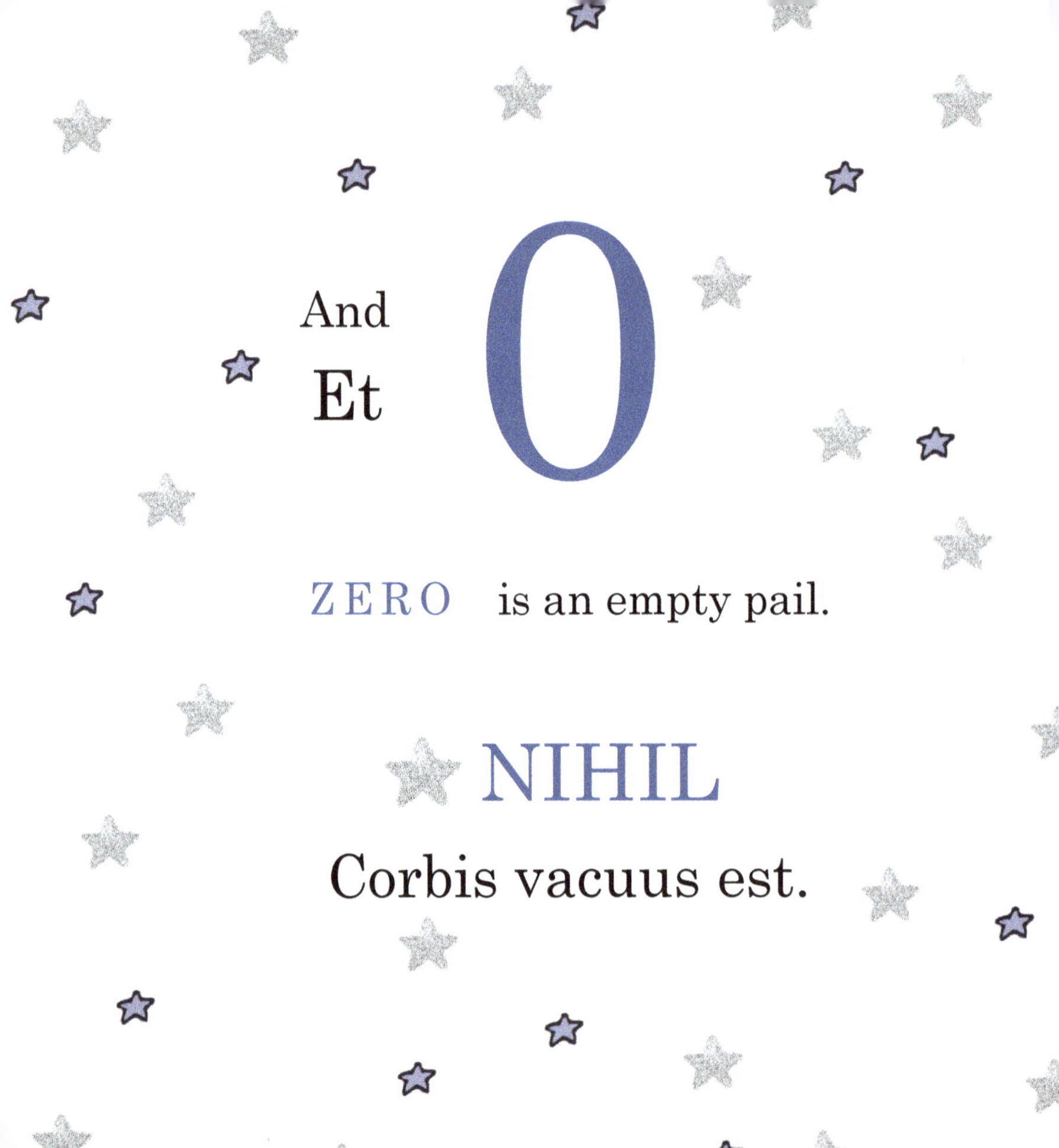

And
Et

0

ZERO is an empty pail.

NIHIL

Corbis vacuus est.

IT'S EMPTY!
VACUUS EST!

Thank you for playing with us today.

We had a lot of fun too!

Gratias ago, quod nobiscum hodie lusisti.

Id nobis quoque placuit!

We are your Number friends,
Zero to Ten,
Who will be here for you~
Sumus amici numerorum tui
ab nihilo ad decem.
Semper pro te stabimus.

Bye-bye now!
See you again soon!
Vale!

The Numbers are *SINGING* too!

To sing-a-long, look for Miss Anna Number Story
at your favorite music store like iTUNES.

MP3

Numbers 0-10
IDENTIFYING
& COUNTING

Numbers 11-20
& Ordinals

first, second, third...

Numbers 0-100
& Place Values

ones, tens, hundreds...

About Clocks
& Telling Time

hours, minutes, second...

Number Story 1 & 2

isbn: 978-0-996216-48-7

Number Story 3 & 4

isbn: 978-1-945977-01-5

Number Story 5 & 6

isbn: 978-1-945977-06-0

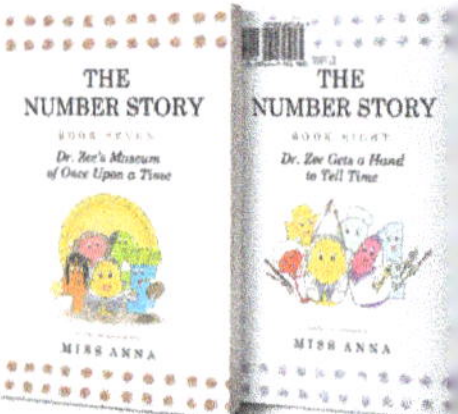

Number Story 7 & 8

isbn: 978-1-949320-40-

For more Miss Anna books to love,
visit us at

w w w . m i s s a n n a b o o k s . c o m

Numbers are working hard all over the world!
Come Travel the World with Us!